Interpersonal Dynamics Consultation
A Manual for Clinicians

Truth springs from argument amongst friends

David Hume

Nothing ever becomes real till it is experienced

John Keats

When two personalities meet, an emotional storm is created

Wilfred Bion

Published and distributed by

The Forensic Education Department and Forensic Psychotherapy Department at West London Mental Health NHS Trust

Interpersonal Dynamics Consultation

A Manual for Clinicians

Gabriel Kirtchuk

John Gordon

Maggie McAlister

David Reiss

First published in 2013 by the Forensic Education Department and the Forensic Psychotherapy Department at West London Mental Health NHS Trust

This book is available from web:-

www.lulu.com

www.amazon.co.uk

Main entry under title:

Interpersonal Dynamics

Includes bibliographical references

ISBN 978-1-4709-7358-2

Table of Contents

Acknowledgements   iii

About the authors   iv

Foreword by Matthias von der Tann   vi

Overview   1

     Introduction   1

     The intrinsic facts of clinical life   1

     The organisational context   3

     Interpersonal dynamics   6

     Transference and countertransference   7

     Importance of interpersonal dynamics (ID) for clinicians   9

     Theoretical and empirical background: mapping relationships   9

     Development of the ID four-perspective consultation   11

     Identifying the negative to strengthen the positive   13

     Administration   14

     Time   15

     Setting   16

     User qualifications   16

     Formulation   16

Table of Contents

Implications for care plans and risk assessment   19

The interpersonal circle – (circumplex)   20

List of items for interpersonal perspective   24

Glossary of cluster items   25-69

Bibliography   70-72

Appendix   73-79

Acknowledgements

The authors are extremely grateful for the inspiration and support from many colleagues which have made this manual possible. In particular, we would like to thank Professors Manfred Cierpka and Gerd Rudolf who, with their associates in Germany, conceived and developed the concept of OPD and came to London to train many of us; Dr Mattias von der Tann and Dr Carine Minne for pioneering the OPD perspective in the UK; Professor Pamela Taylor and Dr Janet Parrot for endorsing this approach within Forensic Psychiatry; professional colleagues who have participated in the Ealing Forensic Psychotherapy Department's Interpersonal Dynamics weekly workshop, especially Dr Richard Ingram and his team in Belfast, Dr Geoffrey Ijomah from Rampton Hospital and Dr Oliver Dale, the Personality Disorder Lead in South CMHT Hammersmith and Fulham; the many members of multi-disciplinary teams who provided us with opportunities to carry out Interpersonal Dynamics Consultations; West London Mental Health NHS Trust management and the Learning and Development Department for their steady encouragement; the National Forensic Psychotherapy Training and Development Strategy which embraced Interpersonal Dynamics; and, the many colleagues who collaborated at various stages of the project in forensic units across the UK.

Finally, we would like to recognise the important contribution made to the production of this manual by Donna Reid, educational administrator of the Forensic Education Department at West London Mental Health NHS Trust.

About the Authors

Dr Gabriel Kirtchuk is a consultant psychiatrist in psychotherapy and a psychoanalyst. He has worked in forensic settings for the last seventeen years and has established a Department of Forensic Psychotherapy, at the Three Bridges, West London Mental Health NHS Trust, Uxbridge Road, Southall, Middlesex, UB1 3EU, UK. He is also the lead clinician of the National Forensic Psychotherapy Training and Development Strategy, as well as the chair of the Special Interest Group in Forensic Psychotherapy at the Royal College of Psychiatrists. He is an honorary clinical senior lecturer at Imperial College London and is a fellow of the British Psychoanalytical Society. His areas of interest have been the development of psychotherapeutic approaches within forensic mental health as well as training and educational programmes in this field.

John Gordon is a psychoanalyst and group analyst. He is honorary senior lecturer at Imperial College Medical School and at Buckinghamshire New University and former consultant adult psychotherapist in the Forensic Psychotherapy Department, West London Mental Health NHS Trust and at The Cassel Hospital.

Maggie McAlister is a Jungian Analyst (SAP) and works as a highly specialised adult psychotherapist within a forensic psychotherapy department within the NHS. She previously worked as a dramatherapist for over fourteen years in general and forensic psychiatry and has published several papers on arts therapies and psychotherapy in the field of adult mental health. She is a senior lecturer for the MSc in Psychotherapeutic Approaches to Mental Health jointly run by Buckinghamshire New University and West London Mental Health NHS Trust.

Dr David Reiss MA MBBChir MPhil DFP FRCPsych FAcadMEd is a consultant forensic psychiatrist at West London Mental Health NHS Trust, and an honorary clinical senior lecturer at Imperial College London. His research examines the interface between clinical forensic psychiatry and public policy, including work on homicide inquiries, stalking, interpersonal dynamics and educational issues. His clinical and educational work focuses on enabling the multidisciplinary team to gain an enhanced understanding of patients, thereby improving care and reducing risk. He is the author of numerous journal papers and co-editor of the book *'Containment in the Community'.*

Foreword

Dr. Matthias von der Tann

This manual describes a milestone in the provision of inpatient care for forensic psychiatric patients as the Interpersonal Dynamics (ID) approach establishes, for the first time in forensic psychiatric work, the subjective experience of staff as part of a diagnostic tool. ID has the capacity to become invaluable when applied by multidisciplinary teams in forensic psychiatry settings. There it can help the teams to make sense of stalemate situations and incidents with patients, providing avenues of a new understanding as well as the possibility of different responses.

After five years of thoroughly structuring their experience gained in clinical consultations with forensic psychiatric staff teams the authors present a tool which hopefully will find widespread distribution and application. Adapted from the inpatient 'staff tuning' with Operationalized Psychodynamic Diagnostics (OPD; Stasch, 2004). ID offers, through systematic use of the interpersonal experience of staff and patients, a psychodynamic understanding of what might lie at the basis of very difficult situations which teams in forensic psychiatry units so regularly have to manage. ID introduces new elements into Axis II of OPD by interpreting the original observable items, thus providing additional aspects to the individual item in a clinical case. This is especially poignant where they propose positive elements in the evaluation as defending against and covering up negative elements. The authors also make use of the concept countertransference in a way that this experience might reflect the externalised (projected) subject-side of an internalised object-relationship, informing therefore about an (early) experience of the patient herself. The full formulation even tries to incorporate the index offence as an expression of a rigid template of relating which had to be established very early. The very common splitting of teams is understood as an externalisation of a defence of the patient in the paranoid-schizoid mode - which becomes apparent also in the more extreme naming of the poles of the affiliation axis. All these elements widen the area which OPD was originally designed for because ID intends to enable a grasp of psychotic and perverse states of mind, which are so common in the psychopathology of patients as well as interactions in forensic psychiatric wards.

It is very remarkable and thanks only to the vision of Gabriel Kirtchuk and the persistent work of David Reiss, together with the other authors, that this development could be pursued up to the point presented here, despite all the difficulties and discontinuities currently facing inpatient forensic psychiatric work in the UK National Health Service (NHS). To start with, in 2005, there were OPD training courses in London (which I had the privilege to organize) conducted by Manfred Cierpka and Gerd Rudolf, key authors of OPD, which were mainly attended by forensic clinicians. Those clinicians – in the first line: Pamela Taylor and Carine Minne – not only felt the need for a new systematic clinical tool, but they also were sufficiently open to get trained in OPD. However, it required Gabriel Kirtchuk's imagination to look for, find and develop a feasible implementation of what he had learned about the original tool into inpatient forensic psychiatric work. Although the second version of OPD, OPD-2 had in the meantime been published, pole 2 of the first OPD version appeared - due to its stronger focus on dysfunctional relationship patterns - to be a more suitable starting point for the development of a specific tool. The further development of ID has also focused on using a language which would make this tool widely useable for every clinician and staff member independent of their training. This extended the original OPD principle not to use any school specific terms.

That this is a handbook for clinicians is obvious from the distribution of text and glossary. It is a guide to systematically making sense of statements about feelings of staff and patients by associating them with items which then can be ordered in a specific way. Those who work with it may therefore use the principles in order to see where to place a statement about a specific feeling. The authors have deliberately withstood the temptation to provide the reader with examples of a full formulation – a wise self-limitation which makes it clear how much this manual is a tool best to be used within the context of a certain understanding and handling of interpersonal relations on a ward.

This is clearly work in progress – and therefore it should not in my view be understood as written in stone, but rather seen as something which is and will be continuously enriched by the further clinical experience of those who use it. In that sense it is also an important part of the ongoing efforts to further develop the forensic module and application of OPD.

Overview

Introduction

In clinical work, an awareness of patients' subjective experiences, particularly their perceptions of interpersonal relationships, is indispensable. The aim of this manual is to improve care and treatment planning by describing and supporting a structured approach to eliciting patients' core relationship patterns. These patterns consist of the roles and scenarios into which they repeatedly cast themselves and others with whom they interact. Maladaptive patterns, in which vicious cycles and self-fulfilling prophecies of misperception, misunderstanding or provocation escalate, cause pain and havoc in personal relationships and can adversely affect both professionals' decisions and the overall delivery of treatment. We are concerned to show how to use vital information that is often not made available to treatment teams in order to understand such potential pitfalls rather than succumb to them.

The intrinsic facts of clinical life

Routine clinical meetings in mental health settings, such as ward rounds, usually involve discussion of patients' symptoms and behaviour. This focus, which derives from the *Diagnostic and Statistical Manual of Mental Disorders IV-RT* (APA, 2000) and the *International Classification of Diseases* 10 (WHO, 1994) frameworks, is essential. However, it does not include the wealth of untapped information embedded in the complex and often subtle interpersonal communications arising from the working relationships between patients and staff. Our premise is that formalised psychiatric evaluation and risk management are immeasurably enhanced by systematic exploration of these interpersonal scenarios, as they emerge in the treatment context.

Modern mental health services are essentially multi-disciplinary and multi-agency. The team approach to caring for and treating patients, based on shared goals, competencies and capabilities, is a central clinical and political imperative. Nevertheless, often opposing this objective is the reality that different professions organise their interventions separately and employ different languages with limited conceptual and clinical overlap, sometimes to the detriment of patient care.

Beyond the effects of continual changes regarding professional roles and identities imposed by professional and governmental policies, there are two additional factors that contribute to a lack of working together in institutional and community care – often with the kinds of catastrophic consequences attributed by formal inquiries to systemic failures of communication: the interpersonal context, and the organisational context. In the first of these, the professions tend to go in different directions as a result of pressures arising from the interpersonal impact of patients' dysfunctional behaviour and communication. There is an active process by which individual members of multi-disciplinary teams and agencies, as well as the separate professional groups themselves, are differentially perceived, used, valued and devalued by patients. When members of one profession, say nurses, feel devalued by patients, compared to doctors who are idealised, it becomes more difficult for the two disciplines to cooperate together.

Patients who have suffered from abuse or neglect in early childhood and perhaps also experienced victimisation as adults may, without realising what they are doing, influence others through the repetition of previous adverse interpersonal patterns. People with these difficulties are found in all areas of psychiatric and social work and are especially prominent in forensic settings. Usually we can recognise them by the strong emotions they are able to generate in staff. For example, some are able to divide the team's views on how they should be managed. Others may elicit intensely positive or negative feelings in specific members of staff. In the extreme, an outsider listening to the various strands of a team case discussion might think that different team members were describing two completely separate patients.

The organisational context

The second factor which can lead to less than optimally coherent service delivery is based on a psychodynamic/systems theory of organisations. All members of staff must struggle to create a picture in their minds of the overall objectives of their organisation and to be aware of how their roles link with each other in order to fit in with these objectives. The work that they have been set up and authorized to do, not by themselves alone but by multiple stakeholders in the organisation's external environment, cannot be accomplished if staff do not have a game plan and the resolve to stick to it. Failure to achieve the objectives – whether converting raw materials into cars for sale or transforming minds acutely disturbed on admission into more stable mental states prepared for discharge – means failure of the organisation.

To stand any chance of completing the job by means of integrating their roles, members of an organization must be in touch with the reality of the work, but in many mental health settings, this reality is extraordinarily painful and frightening. To survive and navigate their everyday encounters with a work reality infused with the fragmented minds of patients, expressed through impact in dysfunctional interpersonal scenarios, individual members of staff deploy their own resources to cope with the aroused anxieties, emotions and impulses. Such individual self-protective strategies are based on the life experiences and personalities of each organisational member and are, consequently, more or less mature. However, organisations, like all groups, 'offer' their members shared or group-level methods of protection to bolster their individualized attempts to shield themselves from the destabilizing emotional glare of direct contact with patients. These shared procedures are group norms, generalized prescriptions on how to carry out the work which may be taught formally, as part of explicit organisational training procedures, or are informally, often implicitly, absorbed as individuals are initiated into the organisation's ways of doing things.

Like individual coping strategies, the group-sanctioned methods of operation may be more or less mature: more or less in touch with the reality of the work. Unless they are conceptualised and subjected to evaluation, such normative working routines, while they may enable staff to survive and remain in their settings, characteristically pull the direction of their energies and efforts away from the primary aims of the organisation. A central finding of research based on this organisational perspective is that the normal, adaptive division of functional roles becomes splintered and fragmented: organisational functioning comes to mirror and reiterate the very disordered states it is meant to transform. Different professional groups may start to concentrate on partial, and therefore 'manageable', aspects of the work; they may even focus exclusively on different aspects of the patient in isolation from the particular focus of other colleagues. Anxiety intrinsic to the reality of facing an integrated task evaporates in the reassuring refrain that patients are 'settled', while managers complain that nurses remained in their offices throughout the shift and nurses complain that doctors and therapists are only on the ward for a limited time. Blame is passed around, along with feelings of inadequacy. Paradoxically, some familiar, almost automatic and 'settled', working practices may increase the possibility of organisational failure, a completely unintended consequence of an understandable and necessary quest for security.

An approach to facing the facts of life in mental health settings

To overcome all of these inescapable forces, we believe it is vital for each clinical team to develop an explicit framework to organize its clinical practice so that gaps in communication can be monitored and addressed. Specifically, the different professionals involved in a patient's care must meet regularly to coordinate their separate views based on certain shared concepts and a common language. By doing this all staff, even if they come from different theoretical backgrounds, can work in partnership to interpret and understand the observed behaviour and symptoms of the patients, as well as their responses to the various treatment interventions.

The method described in this manual is a further development of an approach based on Operationalised Psychodynamic Diagnostics (OPD) which has been described by Stasch (2004). It is able to reveal the underlying dynamics of a patient's interactions in a way that can be comprehended and contributed to by all multi-disciplinary team members. Objectivity, which is seen as the gold standard of scientific measurement, is prized and promoted in contemporary clinical practice, whereas subjective experience, which has traditionally been viewed as not meeting the criterion of detachment, has consequently been marginalised in many forms of psychiatric and psychological discourse. We propose that subjective experience, in particular the ways that staff react to patients and *vice versa*, can be codified, organised and contextualised so as to make it a valid and reliable tool in routine clinical work.

Interpersonal dynamics

Stable, predominantly positive interpersonal relationships are central to mature personality development and good mental health. Unrewarding, painful or hostile relationships are inevitable and, within the context of positive interactions, are not only made safe but contribute to resilience. We usually experience our everyday interactions as a dynamic balance between initiating contacts with others, responding to their approaches, and realizing that we and they are also always considering possible interpretations of each other's initiatives and responses which, according to whether they are positive or negative, feed into the relationship and drive it along. To 'mentalize' (Bateman and Fonagy, 2004) relationships is to consider simultaneously our own states of mind – motives, wishes, intentions, fears, perceptions, initiatives and responses - and those of others with whom we are interacting. This contributes to tolerance, to an awareness of uncertainty and complexity, and to an appreciation of nuance and surprise in human relationships.

Our encounters in the clinical setting, however, are often of a different nature and show particular maladaptive characteristics. First, many of our patients have a *limited range* of ways of relating to others. For example, they may tend to see themselves as only needy and dependent and view others primarily as (adequate or inadequate) caretakers. Second, such relationships tend to be employed *repetitively, rigidly and inappropriately* in a broad range of interactions with others. From a psychodynamic perspective, such inflexible patterns of interpersonal interaction, which may originate in childhood and continue to determine the basic shape of adult functioning, are called transference relationships. A third characteristic of dysfunctional interpersonal interactions is that the patient tends to see himself as merely and perpetually *responding* to the active behaviours of others and pays relatively little attention to how his own approach or impact might affect them. Finally, many of our patients tend to *resort to action* as a response to the perceived motives and behaviour of others, rather than to reflect and try to understand the possible meanings of the interaction. In turn, patients' tendency to act out repeatedly can also impair staff's capacity to think and can lead to counteractions (a form of countertransference) which may include breaking professional boundaries.

In essence, our patients focus actively on their perceptions and interpretations of the manner in which they believe others treat them, as well as reactively responding, primarily by actions and impacts. They are accordingly less able to be aware of their own initiatives and of how these might be interpreted by others. They also do not experience what we have referred to as their active perceptions and interpretations as such; for them these are facts about the other, not products of their own minds which may or may not accurately take account of the reality of others' intentions and behaviour. At its extreme, perceptions and interpretations of self and others' behaviour may take the form of delusional belief. Some of the examples in the glossary include this type of content.

Transference and countertransference

We have alluded to the *transference* as a rigid, compulsory repetition in the present of a relationship pattern which originates in the past, typically in infancy and childhood, and which out of awareness continues to habitually inform and be imposed on current interactions with significant figures in the patient's environment. The *countertransference* refers to the counter-responses of the recipients toward whom the patient's transference(s) are directed. Staff (other) countertransference as responsiveness to the patient includes the whole gamut of human emotional, cognitive and action potential. So patients may 'select' from this available pool just the responses which are required to fit their internal scenarios and produce dramatic re-enactments with current others. Of course others are also primed by their own transference patterns, as well as by the understandable effects on them of their patients' impacts, to respond either in accordance with patients' transferences or to resist. In this way, transference-countertransference can be seen as 'schemas', the lifelong accumulation of relationship experience in the form of cognitive affective schemata (Horowitz, 1991) and therefore we believe this approach can lend itself to different modalities, like schema focused therapy.

Transference-countertransference configurations enacted between each patient and those members of the staff involved in his treatment often evoke a complementary self relating to other (Racker, 1968) response. For example, the patient plays the role of self, say an appreciative little boy, with one member of staff who responds in the role of a satisfied mother. With another member of staff, the patient in the role of an angry child, transfers an attitude of hostile rejection; and the professional feels like a useless mother. With yet a third, the patient repeats the self role (experience) of feeling bad and guilty, while the other is induced to be a critical, rejecting mother.

Roles can be - and often are - reversed when the patient plays the part of '(m)other' while staff are put in the position of the patient's self within the particular transferred scene. It can be very difficult for staff to be cast by the patient in more 'negative' roles, for example, when the patient is really pushing us to respond critically or making us feel completely worthless and useless. But it is even harder when the patient, in reversing the roles, succeeds in making us experience what it was like to be him or her as a child. This can be exceedingly painful, may feel cruel, and can lead to staff attempts to evade the experience. On the other hand, exploring the countertransference offers an exceptional opportunity to get to know the patient from the inside and can enable alternative responses that are less maladaptive and more concordant with the real treatment needs of the patient.

Importance of interpersonal dynamics (ID) for clinicians

It is possible, within a multi-disciplinary team (MDT) setting, through systematic examination of the information available about an individual patient, to delineate the patterns of dysfunctional interpersonal interactions starting in childhood and continuing in the staff-patient interactions on the ward. Identifying the dysfunctional interpersonal interactions may inform risk assessment and allows the staff to devise specific care plans to help modify the dysfunctional interactions.

Through experience, we know that various members of the MDT highlight differing aspects of the patient which may not be known to colleagues. This sharing of multiple pictures of a patient's personality and behaviour as expressed in their relationships adds to the coherent understanding of the patient by the staff.

A discussion of the staff-patient interactions can, therefore, improve communication within the staff group through provision of a common understanding of the dysfunctional interactions and lead to a more consistent response to patients' behaviour. It can also identify situations in which staff may perhaps have acted out in response to patients' behaviour and enhance the ability of staff to think and reflect. This is turn would reduce the risk that staff would breach professional boundaries.

Theoretical and empirical background: mapping relationships

Birtchnell (1993) described how relating, an activity which is universal across the animal kingdom, confers advantages upon those who are successfully able to engage in it. He outlined how relationships could be described by two axes: proximity (horizontal) and power (vertical). Freud (1950) had previously proposed that there were two types of human instinct: sexual and destructive, which corresponded to the horizontal axis, and power relationships, represented by the vertical axis.

Freud's work was taken forward by Sullivan (1953), a psychiatrist and psychoanalyst, who understood that infants need emotional contact with others and that early perceptions of interpersonal interactions, initially with parents, greatly influence adult personality development. His theory recognised that everyone fundamentally requires love and power in order to be secure and free from anxiety. He also emphasised how important it is for mental health professionals to be able to understand the world as the patient sees it. The 'object relations' school of psychoanalytic theory further developed this strand of work and highlighted the paramount importance of attachment (Bowlby 1969) and autonomy for normal development (Greenburg and Mitchell,1983).

Bowlby's original concept has been substantially developed by those working in the attachment field to produce a comprehensive and in-depth understanding of attachment behaviours and difficulties in interpersonal relationships. Recent developments in neuroscience (Schore, 2001) have explored how brain and personality development are impaired in infants who experience psychological neglect from primary care givers. Fonagy and Bateman (2004) have conceptualised this impairment as a deficit in the capacity to mentalize: to represent mental states. The mentalization based approach is highly relevant to interpersonal dynamics as the aim is to increase the capacity for reflective function in order to better understand the emotions, intentions and beliefs of others, and to differentiate these from those of oneself.

Murray (1938) produced a list of human needs which he saw as themes underlying human behaviour. Leary (1957) conducted empirical research and arranged a selection of Murray's needs around two perpendicular axes (love/ hate and dominate/submit) to form the basis of the interpersonal circle. This arrangement, which describes a spectrum of possible interactions that can occur within a relationship, is also known as a circumplex (Guttman, 1996). Schaefer (1965) proposed a vertical axis, modelled on parenting behaviours, which is defined in terms of allowing autonomy versus control.

Allport (1937: 48) defined personality as 'the dynamic organisation within the individual of those psychophysical systems that determine his unique adjustment to his environment', and he used the term 'dynamics' to refer to an individual's goals and purposes. Benjamin (1996), drawing on both personality theory and the interpersonal circle, produced the Structural Analysis of Social Behaviour (SASB) model. She wanted to develop a more objective understanding of psychopathology in interpersonal terms and demonstrated distinctive relationship profiles for different types of personality disorder. In the forensic context, Blackburn (1998) has used the interpersonal circle to examine the relationship between interpersonal style and criminality in both mentally ill offenders and those with personality disorder. He found that offenders with extensive criminal careers have a more dominant and coercive interpersonal style.

Development of the ID four-perspective consultation

The Operationalised Psychodynamic Diagnostics (OPD) Task Force (2001) have formulated a reliable empirical method to determine stable but dysfunctional patterns in relationships. It took up the circumplex model, was heavily influenced by the SASB (Benjamin, 1996) and elaborated this approach by taking into account the work of others who have also outlined rigorous methods of observing interpersonal interactions (Strupp & Binder, 1984; Weiss & Sampson, 1986; Hoffman & Gill, 1988; Luborsky & Crits-Christoph, 1990;and Horowitz, 1991). The totality of OPD is a comprehensive, validated, structured assessment protocol which combines descriptive phenomenological diagnostics (ICD-10, 1994;

DSM-IV, 1980) with psychodynamic features derived from psychoanalysis. (For reliability and validity data see von der Thann et al., 2007). Patients are assessed on five axes: (I) experiences of illness and prerequisites for treatment; (II) interpersonal relations; (III) conflicts; (IV) structure; and (V) mental and psychosomatic disorders.

In our work with the multidisciplinary team, we concentrate on the framework of Axis II, which concerns the patterns of relationship that characteristically describe patient-staff interactions. Our basic procedure is to conceptualise (adapted from OPD) four perspectives of the patient's core relationship patterns by clarifying (1) how the patient characteristically perceives the other; (2) how the patient responsively experiences himself; (3) how the other (including staff members) usually perceive the patient; and (4) how others/staff experience themselves in their interactions with the patient.

The four interpersonal perspectives:

> Perspective A: The patient repeatedly perceives others so that they are ... (focus is on the other – active)
>
> Perspective B: The patient regularly experiences himself/herself as ... (focus is on the self – reactive)
>
> Perspective C: Others, the staff included, repeatedly perceive the patient as ... (focus is on the other – active)
>
> Perspective D: Others, the staff included, regularly experience themselves as ... (focus is on the self - reactive)

This framework concerns the transference-countertransference configurations enacted between each patient and those members of staff involved in their treatment as experienced in the care setting (Kirtchuk et. al., 2008).

To summarise this key point: both participants in any interpersonal, self-other, interaction will have two basic types of experience. Each has an experience which consists of various perceptions of how the other is relating to the self: an active focus on the other's actions, behaviour, attitudes and states of mind. In turn, as a response to these perceptions of the other, each person also has an experience of him/herself: a reactive focus on the self's actions, behaviour, attitudes and states of mind. For example, to a perception of the other as critical and humiliating, the self may respond by feeling despair or, alternatively, by becoming offended or even running away. The whole forms a reliable and valid empirical structure to determine stable but dysfunctional patterns in relationships (Cierpka et. al., 2007)

An in-depth exploration of these patterns helps to uncover the central, core interpersonal dynamics of the patient which are repeated time and again. Once the core dynamics are identified, it becomes possible to link past experiences in the patient's life with ways in which staff members may be unwittingly caught up in dramatic repetitions of those past patterns. From this material it is possible to hypothesise the significant internalised object relationships of the patient, the internal map of expected, desired and feared relationships based on cumulative experience, and the manner in which these scenarios may influence external relationships on the ward. It is this largely unconscious pattern of relating, the transference-countertransference configuration, that is the focus of the ID four perspective consultation (as described by Stasch et al., 2002).

Identifying the negative to strengthen the positive

We have developed the ID protocol from the first version of the OPD Axis II (OPD Task Force, 2001) in order to specifically focus on the level of pathology and destructiveness of the group of patients with whom we work. In this sense, all of the cluster and item descriptions used in the consultation to elicit interpersonal patterns, including positive qualities such as 'affirming' and 'protecting', can be seen to have a latent, more negative dimension: for example, collusive, biased 'affirming' and godfather-style 'protecting'. Our intention in interpreting the protocol in this way is not to apportion blame or ignore the real strengths of the patients and staff, but to use the consultation to identify what is maladaptive and problematic. This arises from our experience in a forensic setting where professionals are working on the edge of what is emotionally tolerable, where contact with catastrophic narratives is commonplace, and where 'communication by impact' (Casement, 1985) is the rule. In order to grasp the nature of clinical interactions in such an extreme setting and in work with such severely ill and dangerous patients, we have also used language from the Kleinian psychoanalytic perspective to modify some of the terms used in our version of the interpersonal circle (circumplex) and on the protocol. This language (for example, 'destroying' instead of the more usual 'hating', and 'idealising' instead of 'loving') fully encompasses an understanding of the challenges posed by the combination of psychosis, perversion and psychopathic elements faced particularly but not exclusively by forensic mental health professionals. The aim of the consultation is always to elucidate the negative aspects of relating in order to strengthen the positive aspects.

Administration

The interpersonal dynamics of any particular patient should be assessed at a multidisciplinary team meeting. Participants should include staff members who are personally involved with the patient, as well as a facilitator from outside the team who has expertise in formulating core relationship patterns. It is important that experiences are shared because, as we have outlined above, the patient may feel and behave differently towards various carers. The process is non-hierarchical as all perceptions and experiences are valid.

The consultation itself comprises four stages. The initial part is a presentation of background information, based on reports of the patient's past and current significant relationships, offending pathway and index offence, before moving on to consider current relationship patterns with others and staff members. The reasons for the ID consultation are elicited at this stage in order to bring into focus the main problems inhibiting treatment and progress. Next, the interpersonal dynamics are determined using the four-perspective protocol. For each perspective the three or four most salient items are identified. If there are two items identified in one cluster, it is best to choose the most accurate item. However if both items are equally present they should both be chosen, as this indicates the very strong relevance of the cluster. Regarding the first two perspectives (patient's perspective on others and on self), the clinical team must provide clear examples from statements the patient has made in the first person. Alternatively or additionally, a team member may conduct a consultation to the patient in order to obtain his/ her views on the first two perspectives and offer an opportunity for the patient to contribute to his/ her own recovery and overall treatment. All items that are selected on the protocol must have clear evidence, based on statements and examples. A formulation is then produced which attempts to link the emerging current interpersonal dynamics with past relationship patterns, including the index offence. Finally the treatment strategy is reviewed in the light of the ID formulation.

Time

We currently undertake this process in two meetings of up to one hour each. In the first meeting the first three stages take place; between the meetings the formulation is written up; and the second meeting is for the review of clinical care. However, more complex cases may need additional meetings to understand the implications of the formulation. The ID consultation process may be repeated every six months, usually before the patient's care planning meeting.

Setting

It is strongly recommended that this instrument be used in a group setting involving various members of the MDT. However this instrument can also be used by the individual practitioner to enhance the over-all assessment in routine clinical work, although obviously the views of the team members won't be included in this approach. In our experience the presence of professionals from three or more disciplines within an MDT such as doctors, nurses, arts therapist, occupational therapist, psychologist or social worker is most useful in collecting a sufficient amount of clinical information related to staff-patient interactions in order to identify the most relevant patterns of dysfunctional interpersonal relationships.

User qualifications

The ID Consultation is designed to help all mental health professionals working within multidisciplinary teams. It is especially recommended for those who have continuing contact with patients and who make routine mental state assessments within a clinical setting, as well as trainees who are specialising in a relevant field. It is important that senior clinicians take a lead in organising ID meetings and arranging for background information on the referred patient to be available.

It is also recommended that MDT members regularly utilise the protocol within their setting in order to obtain a high degree of proficiency in arriving at a formulation which can contribute to care plans and risk assessment.

Formulation

The formulation should consist of two separate yet linked narrative strands. A basic formulation uses information on the four perspectives derived from the highlighted clusters and items on the protocol which have emerged in the consultation. Therefore, the formulation begins with A) the patient repeatedly perceives others as ... B) and in response, the patient experiences him/her self as ... continuing on to C) others repeatedly perceive the patient as ... D) and in response experience themselves as ... So for example, the patient repeatedly perceives others as either domineering and imposing, or as abandoning and responds by believing he/she knows best and cutting off contact. Consequently, others (the staff) repeatedly perceive the patient as insisting on his position and keeping up a barrier which leads them to want to dominate and impose on him or give up in despair. In effect this simply reinforces the patient's original perception that others are controlling and abandoning.

In collecting evidence for each perspective, it is important to identify features in both the here and now relationships within the clinical setting as well as in past experiences of the patient. The four perspectives often describe a dysfunctional cycle that repeats relationship patterns from past to present (see figure 1 below). It is important to re-emphasise that the consultation is concerned with these dysfunctional aspects of relating because, although they are precisely the ones which are most difficult to bear, to attend to and to think about, they also contain very significant information about the characteristic functioning of the patient's mind as expressed in his social relationships.

(Figure1)

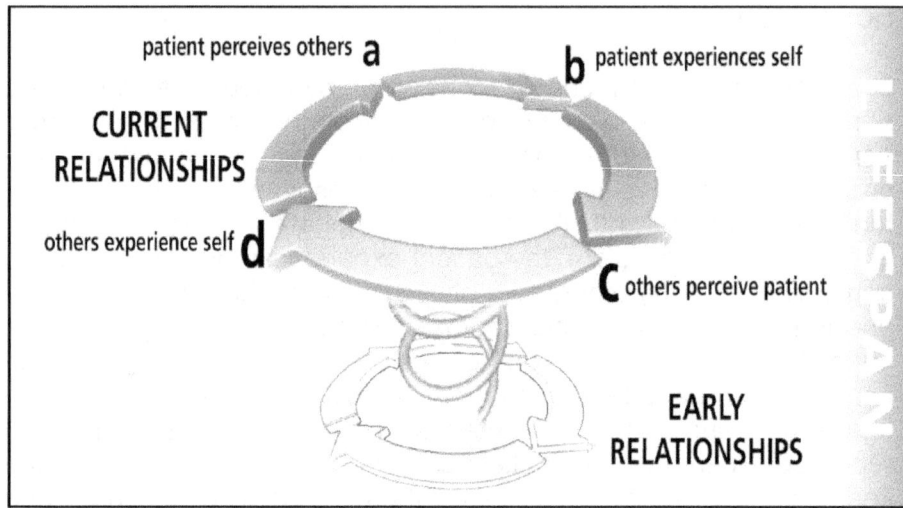

A subsequent, expanded formulation includes a more detailed emotional narrative of the patient's core repertoire of relationship patterns as they have developed in the course of the patient's life, especially with significant early caretakers as illustrated in the diagram above. The presence of an appropriately trained practitioner will significantly facilitate the construction of this more comprehensive formulation. The key task is to show how the patient's perceptions of others may involve misattributions of characteristics belonging either to the patient or to important figures in the patient's past, who in turn respond automatically to the impact of these attributions. For example a patient who cannot tolerate feelings of intense guilt may behave in such a way so as to aggravate and so draw criticisms from staff, who accordingly take on the role perhaps of a highly critical parent. Subsequently, the patient will naturally feel guilty for upsetting staff or rebel against unjustifiable criticism which might make staff even more authoritarian. In this manner the patient's own guilt seems to be dramatised in the form of rebukes, now emanating from a person in the external world. Conversely, staff may also contribute to the dysfunctional relationship as a result of their own make up or other unresolved conflicts, e.g. conflict within the staff team. An important aim of the expanded formulation is to illuminate the presenting interpersonal problems which initiated the consultation and those which might increase risk, including the interpersonal dynamics represented by the index offence. The formulation is accessible to all professionals and provides a framework within which each professional can act in accordance with his or her own way of working. However, the process of discussion itself is as important as the formulation.

Implications for care plans and risk assessment

An important aim of the consultation should be to allow the formulation to inform future care plans, risk assessment, and the overall treatment and management of the patient. The implications of the core interpersonal dynamics may cast light on how best to manage the patient's care. For example, one may consider whether, due to a tendency to perceive the patient as 'special', one is caught up in making allowances that further the patient's grandiosity or feelings of entitlement. A practical consideration would be to review any special arrangements in place for that patient. Another example may be where a patient is allowed too much independence because of a difficulty in managing the level of despair and hopelessness the patient engenders in staff. In this sense the staff may be enacting an earlier relationship where caregivers effectively abandoned the patient and left him to fend for himself. All of these situations may have an impact on risk of further offending, or re-create scenarios that are similar to the index offence and so need careful consideration and management.

The interpersonal circle (circumplex)

The interpersonal circumplex is a two dimensional visual representation, a circle drawn around horizontal and vertical axes, which allows one to plot the core relationship interactions between a subject and the significant others in his or her life (Figure 2). As mentioned above in the theoretical and empirical review, the horizontal axis traditionally denotes a range from extremely negative to extremely positive positions in regard to proximity/affiliation ('destroying'/'idealising'); and the vertical axis similarly charts a spectrum of positions in regard to control ('allowing independence'/'controlling'). The circle can also be viewed as composed of four quadrants, marked by the relationship characteristics (clusters) indicated at right angles. At the mid-point of each quadrant, along the perimeter of the circle, further descriptions of interpersonal interactions may be placed which are blends of the two clusters located at the quadrant's boundaries. For example, in the lower left quadrant of the active circle below, bounded by 'controlling' and 'destroying', 'blaming' is located at the intermediate point on the basis that 'blaming' contains modified aspects of both 'controlling' and 'destroying' clusters. The quadrant structure will be described in more detail below.

Figure 2: The Circumplex

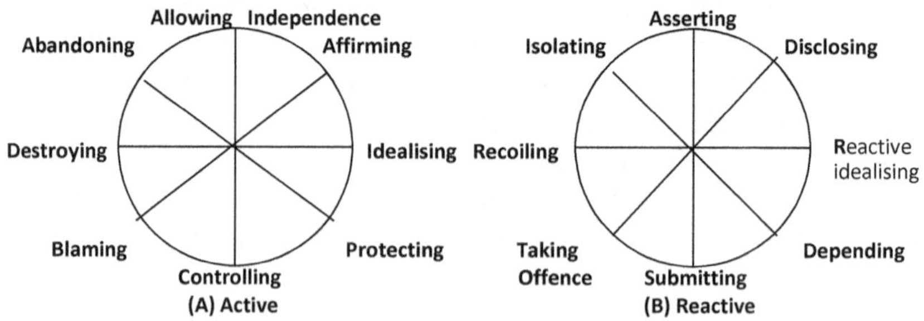

20

Although two circles are illustrated , they should be understood as showing two layers of the circumplex map, **active** and **reactive** (Benjamin 1996), which together portray different aspects of the interpersonal relationship. As mentioned, the active layer always refers to how the subject, whether patient or staff member, actively perceives and interprets the actions, behaviours, attitudes and states of mind of the other. The reactive layer refers to how the subject accordingly responds to these perceptions.

The active circle 1-16

The vertical axis of the active circle is 'allowing independence' versus 'controlling', and the horizontal axis is 'destroying' versus 'idealising'. Starting from the top of the circle, these four points mark key stages in a journey through the clusters divided into four quadrants. The first quadrant ranges from 'allowing independence' to 'idealising' with 'affirming' at the midpoint. The offer of freedom perceived in the 'allowing independence' cluster may amount to a neglectful ignoring ('treating as self-sufficiently independent') which echoes the 'destroying' – 'abandoning' clusters in quadrant 4 to its left; or, moving clockwise, it may become a collusive indulgence expressing an extremely accepting and positive view. This becomes 'idealising' when the other is viewed as treating the subject as special and exceptional, leading to an over-estimation and inability to see any negative aspects. The second quadrant moves from 'idealising' to 'controlling' by way of 'protecting', The milder form of 'protecting', closer to 'idealising', is expressed as 'attending to and caring for in every way' ; whereas the form influenced more by the 'controlling' cluster at the bottom of the quadrant is described as a mafia "godfather"-like 'instructing and patronising'. The second quadrant ends with 'controlling' which reflects varied facets of ruthless domination . The third quadrant moves from 'controlling' to 'destroying' with 'blaming' at the mid-point, the latter expressed first as accusations in a more 'controlling' mode and then in behaviour which becomes more destructively 'putting down and humiliating'. At its extreme the third quadrant becomes the 'intimidating and 'attacking' behaviour of the 'destroying' cluster. The fourth quadrant moves from 'destroying' to 'allowing independence' with 'abandoning' situated at mid-point. The second item in the 'destroying' cluster is 'rejecting and excluding', a more subtle form of 'destroying' contact which shares aspects of the adjacent 'abandoning' cluster . 'Abandoning' then leads us back to the top of the circle, returning to the neglectful freedom of 'allowing independence' with which we began.

There are eight clusters each of which includes two descriptive items. It may seem that there is little differentiation between the two items, but as just indicated by the examples characterising our journey around the perimeter of the active circumplex, they are designed to denote subtle differences influenced by their positions in relation to the four key points on the axes, as well as to neighbouring clusters at the midpoints. Once again focusing on the first cluster, the perception of 'allowing independence' might be closer to 'abandoning' (and the 'destroying' cluster of quadrant 4) or, alternatively, to 'affirming'   (and the 'idealising' cluster of quadrant 1).

The reactive circle 17-32

The vertical axis of the reactive circle is 'asserting' versus 'submitting', and the horizontal axis is 'recoiling' versus 'reactive idealising'. The first quadrant moves from 'asserting' to 'reactive idealising' with 'disclosing' at midpoint. 'Asserting' might be expressed through 'defying and opposing' behaviour which, by keeping the responding self distant from others, is coloured by the 'isolating'/'recoiling' clusters characterising quadrant 4 on its left; or, moving clockwise towards the 'disclosing' and 'reactive idealising' clusters of quadrant 1, self responds by 'insisting on his/her position', imbued with 'over revealing and intrusive' behaviour and increasing over involvement and merging with the idealised other. The second quadrant covers 'reactive idealising' to 'submitting' with 'depending' at the midpoint. 'Depending' is expressed by 'over relying on' as it approaches 'reactive idealising' of the other and, changing its tone, becomes increasingly parasitical and passive as it nears the 'submitting' cluster at the bottom of quadrant 2. The third quadrant, moving from 'submitting' to 'recoiling', represents how "the worm turns". From abject submission the self begins to develop a 'hurt and touchy', increasingly 'indignant and self-justifying' stance, which ultimately leads to 'recoiling' from the other. 'Recoiling' condenses alternating fight/flight responses: like a recoiling rifle, movement can be aggressively toward ('showing disgust') or away from ('running away') the other. Finally, the fourth quadrant begins with the 'recoiling' cluster, continues to 'isolating' at midpoint and completes the circle with 'asserting'. Bordering the flight aspect of the 'recoiling' cluster, 'isolating' is expressed by 'running away'. As it approaches the top of quadrant 4 , 'isolating' begins to reflect aspects of 'cutting off contact', the 'defying and opposing' of the 'asserting' cluster with which our journey started.

As in the active circle, each of the eight reactive clusters comprise two items which denote subtle differences influenced by their positions in relation to the four key clusters on the axes and to adjacent midpoint clusters. Therefore, the 'asserting' response at the top of the circumplex may be closer to 'keeping up a barrier' as expressed by the 'isolating' - 'recoiling' clusters of the preceding fourth quadrant; or, alternatively, 'asserting' becomes more insistent and absorbing as it approaches the 'disclosing' -'reactive idealising' clusters of the first quadrant.

List of items for interpersonal perspectives

Active circle

1 Treating him/her as self-sufficiently independent

2 Behaving as though he/she knows best

3 Supporting and agreeing

4 Accepting and admiring

5 Treating him/her as special

6 Idealising him/her

7 Attending to and caring for him/her in every way

8 Instructing and patronising him/her

9 Domineering and imposing on him/her

10 Manipulating and exploiting

11 Accusing him/her

12 Putting down and humiliating him/her

13 Intimidating and attacking him/her

14 Rejecting and excluding him/her

15 Deserting him/her

16 Ignoring him/her

Reactive circle

17 Defying and opposing him/her

18 Insisting on his/her position

19 Over revealing and intrusive

20 Pouring out concerns and anxieties

21 Over involved

22 Over sympathetic

23 Over relying on him/her

24 Draining him/her

25 Appeasing and complying with him/her

26 Giving up in despair

27 Indignant and self-justifying

28 Hurt and touchy

29 Running away from him/her

30 Showing disgust toward him/her

31 Cutting off contact

32 Keeping up a barrier

Cautionary Note

In selecting the most applicable items from the tuning sheet, scorers must bear in mind that some of the items require a certain amount of self awareness and insight on the part of the patient and / or others. If this insight is absent, then you may need to consider whether the item is more appropriately categorised on a different perspective. Since it is easy to stray into the wrong perspective when considering items, the facilitator of the consultation should pay special attention to this. Discussion of these issues, however, is a useful and helpful part of the process.

Glossary of cluster Items

The following glossary provides a brief description of each cluster with examples of the two items from each interpersonal perspective. The figure illustrates where the cluster is situated on the interpersonal circle with the neighbouring clusters also indicated.

Allowing Independence

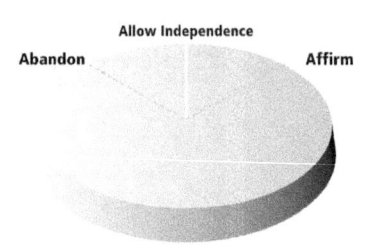

This cluster is between the abandonment and affirming clusters; therefore, there may be elements of these in the types of independence that patients believe they are permitted. Items on this cluster may be particularly relevant for patients with a history of neglect and abandonment when they were left to fend for themselves as children. The autonomy and independence perceived by the patient may be an enactment of an earlier relationship in which the patient was either seen as being more capable than was really the case or attributed with qualities he/she didn't really possess. The staff team may enact this by treating the patient as less unwell than he/she really is or as going along with the patient's wish to be left alone when this isn't appropriate.

Perspective A

Item 1.   The patient repeatedly perceives others as *treating him/her as self-sufficient.* 'You know I can look after myself', 'They know I can cope'. 'When my father died my mother told me I was the man of the house.'

Item 2.   The patient repeatedly perceives others as *behaving as though he/she knows best.* 'You let me decide which groups to go to.' 'My mother thought I was a genius.' 'My mother let me decide everything for the family.' 'You let me do what I want.'

Perspective B

Item 1.   The patient repeatedly experiences himself/ herself as *self-sufficiently independent.* The patient believes he can manage on his/ her own and needs no-one. 'I don't need you.' 'I taught myself at school.'

Item 2.   The patient repeatedly experiences him/herself as *behaving as though he/she knows best* e.g. 'I know what's best for me', 'I'm going to do it my way.' 'Teachers could never teach me anything.' 'I always understood life better than my parents did.'

Perspective C

Item 1.   Others repeatedly perceive the patient as *self-sufficiently independent.* Others see the patient as a law unto him/ herself. 'As a child he always did his own thing.' 'He acts as though he doesn't need anybody.'

Item 2.   Others repeatedly perceive the patient as *behaving as though he/she knows best.* Others perceive the patient as dismissing the views of others. The patient adopts a 'know it all' style. He or she makes up his or her mind without seeking others' advice which is considered to be obviously unnecessary. 'He never came to seek help.' 'My opinion means nothing to him.'

Perspective D

Item 1.   Others repeatedly experience themselves as *treating him/her as self sufficiently independent.* On reflection others may find that they allow the patient to go his/ her own way beyond the norm of usual practice. 'He can have special leave arrangements.' 'I often feel I don't need to supervise him.' 'His mother said that she realised that she let him come and go as he pleased.'

Item 2.   Others repeatedly experience themselves as *behaving as though he/she knows best.* Others may find they give way to the patient's viewpoint and easily lose contact with their own professional judgement. 'We let him decide his own medication.' 'He seemed so knowledgeable about the course that I didn't even consider the possible risks.'

Affirming

This cluster refers to perceptions and offers of support and acceptance which may border on indulgence and collusion. Positive feelings predominate which may mask negative feelings. There is a permissive flavour to this cluster and an inability to

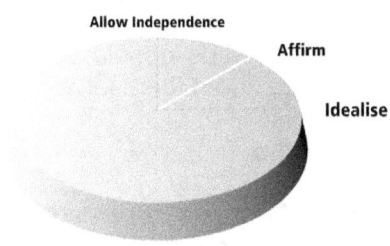

recognise real limitations or obstacles especially when this ensures that any potential conflict is minimised, denied and/ or glossed over. The cluster is located between allowing independence and idealising; therefore, there may be elements of excessive emancipation and idealisation in the affirmation.

Perspective A

Item 3.   The patient repeatedly perceives others so that they are *supporting and agreeing.* 'People tend to go along with my own ideas.' 'My psychologist always sees my side of things.' 'Most of the patients support me in the community meetings.'

Item 4.   The patient repeatedly perceives others as *accepting and admiring* of him/her. 'They always thought my ideas were right.' 'My doctor is always telling me how well I'm doing.' 'My mother always went along with me.'

Perspective B

Item 3.   The patient repeatedly experiences him/herself as *supporting and agreeing.* 'I always agree with my team.' 'I'm always looking out for other patients.' 'When I was little I was always looking after my mum.' 'I look out for my family and make sure they do alright.'

Item 4.   The patient regularly experiences self as *accepting and admiring*. 'I always trust my doctor.' 'I'm the kind of guy who keeps people sweet.' 'I make people feel relaxed and good about themselves.'

Perspective C

Item 3.   Others repeatedly perceive the patient as *supporting and agreeing*. 'He's always helpful on the ward.' 'His father sees him as always willing to lend a hand.'

Item 4.   Others repeatedly perceive the patient as *accepting and admiring*. 'He's always warm and welcoming of me.' 'He's very pleasant and complimentary towards me.' 'She always looks up to me.'

Perspective D

Item 3.   Others repeatedly experience themselves as *supporting and agreeing*. 'I feel like I'm the only one who really understands him.' 'It wasn't really his fault; he just got pulled into things.' 'He was in the wrong place at the wrong time.' 'She was acting in self defence.'

Item 4.   Others repeatedly experience themselves as *accepting and admiring*. Others have a tendency to feel permissive in their dealings with the patient, or many members of the MDT may want to work with him. 'I especially enjoy working with him.' 'I always look forward to my sessions with him.' 'He was his mother's favourite.'

Idealising

The idealisation referred to in this cluster concerns an unrealistic view of the patient or others' real attributes. These may be over-estimated and seen as more positive then they really are due to the lack of ability to acknowledge a more depressing or negative

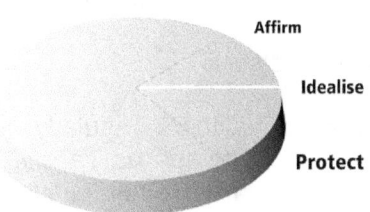

reality. Being seen as special may mean being singled out in either a grandiosely positive or negative way. The cluster is located between the affirming and protecting clusters; therefore, the idealisation may have elements of affirmation and protection.

Perspective A

Item 5.   The patient repeatedly perceives others as *treating him/her as special.* The patient may believe he/ she is allowed special privileges and is given special treatment from other patients. This may take the form of either excessively good or bad treatment. The point is that the patient feels singled out in a special way. 'Staff say I don't belong here because the other patients are all mad.' 'My mother thinks I am a genius.'

Item 6.   The patient repeatedly perceives others as *idealising him/her.* 'All women are in love with me.' 'Every-one wants to have sex with me.' 'When I was five I cooked all the family meals.'

Perspective B

Item 5.   The patient regularly experiences him/herself as *treating him/herself as special.* 'I'm a prophet.' 'I need to save all the children in the world.'

Item 6.   The patient regularly experiences him/herself as (self) *idealising.* 'I'm fabulously wealthy; I run a multi-million pound business.' 'I can have any woman in the world.'

Perspective C

Item 5.   Others regularly experience the patient as *treating him/ herself as special.* The patient may single him/herself out from other patients, see him/herself as different, e.g., not mentally ill. The patient may take on a particular role on the ward where he/she feels entitled to special privileges and treatment. 'She always expects to be treated differently.' 'He always had a sense of entitlement when he was a child.'

Item 6.   Others regularly perceive the patient as (self) *idealising.* Others experience the patient as grandiose, with delusional exaggeration of his/her uniqueness and significance. This may take the form of delusional beliefs or in how he/she expects to be treated. Others perceive the patient as having an unrealistic view of what he/she is capable of. 'He thinks he can save the world.' 'He thinks he's going to make millions of pounds.' 'He says that he and his father worked together in an international company they created for many years before he came to this hospital (when objective evidence points to the contrary).

Perspective D

Item 5.   Others regularly experience themselves as *treating him/her as special.* 'He's different from the other patients. He doesn't really belong here.' 'I've never come across a trickier patient'.

Item 6.   Others regularly experience themselves as *idealising him/ her.* 'With his intelligence he just needs to study English and he'd be alright.' 'He shouldn't be here for long, this place isn't right for him.' 'I wish I could take him home and look after him.'

Protecting

This cluster is located between the idealising cluster and the controlling cluster; therefore perceived or responsive 'protecting' may have elements of idealisation or control. In the more caring aspects, the protection may

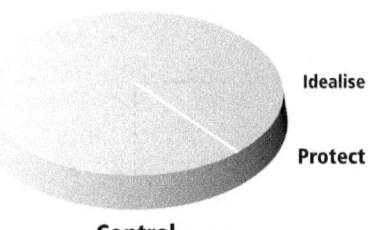

Idealise

Protect

Control

result from an idealised view the patient has of him/herself, where he/she may feel like the only rescuer/ hope for others. The controlling element may come in where the protection is in the service of manipulating the other.

Perspective A

Item 7.   The patient repeatedly perceives others as *attending to and caring for him/her* in every way. This may come across as idealised care, to the exclusion or denial of more negative/ hostile feelings. This item is often present where the patient has split the staff group, and some members are perceived to be all good. 'My primary nurse really understands me.' 'My family gave me everything I wanted' (when objective evidence may point to the contrary).

Item 8.   The patient repeatedly perceives others *as instructing and patronising him/her.* Patient may feel that others protect him/ her in a way that he/she begins to feel 'managed'. He/she may feel cared for in a way that makes him/her feel infantilised or not allowed freedom to make his/her own choices. 'You're always telling me how to do things.' 'I don't need an OT to tell me how to fry an egg.' 'I know how to tie my shoelaces.'

Perspective B

Item 7.   The patient repeatedly experiences himself as *attending to and caring*. The patient sees him/herself as attending to and caring for others. The patient feels that he/she rescues others from difficult situations or provides financial help. 'I look after everybody here.' Some patients can see their offences in delusional terms as caring e.g. a patient who has killed someone may believe they were helping them, in some way.

Item 8.   The patient repeatedly experiences him/herself as *instructing/ (patronising)* 'I need to show her what to do or she'll get into trouble.' 'She doesn't know how it works round here.' 'You doctors don't understand the medication like I do.'

Perspective C

Item 7.   Others repeatedly perceive the patient as *attending to and caring*. This has the quality of idealised care where it may be difficult to see more destructive or aggressive aspects of the patient. Others may only see the caring side of the patient at the expense of other treatment needs. Alternatively others may recognise the patient as becoming the 'mother' for others based on identification with the absent mother; the deeply neglected patient who becomes the super protective carer, overtly towards others but endlessly attempting to look after himself or herself. 'He's a great help on the ward.' 'She always looks after other patients.'

Item 8.   Others repeatedly perceive the patient as *instructing and patronising*. 'She was always trying to tell me what to do.' 'He always talks down to me in our sessions, as if I'm a child.' 'He always tries to take charge of the community meeting, but in quite a nice way.'

Perspective D

Item 7.   Others repeatedly experience themselves as *attending to and caring*. 'She just needs lots of care and understanding … and I'm the person to provide it'. 'I always give him what he asks for.'

Item 8.   Others repeatedly experience themselves as *instructing and patronising*. 'I seem to end up telling him what to do.' 'I keep treating her like a child.' 'I agreed to change the medication because I felt I was patronising him.'

Controlling

This cluster concerns being controlled in a very direct way. This may take the form of endless demands or of feeling one is being taken over or hijacked by the other. This cluster is relevant where there is a very limited scope to have an independent view or the opportunity to act freely, and

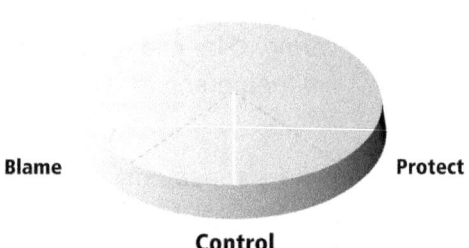

the result is a restriction in making decisions or in taking action. A patient may find a separate view unbearable or staff may find it unbearable to present a separate point of view to the patient. Psychopathic personalities may have items on this cluster.

Perspective A

Item 9.   The patient repeatedly perceives others as *domineering and imposing on him/her.* Any attempts to care for the patient or make contact may be experienced by the patient as an attempt to dominate and control him/her. Patient feels massively intruded into and experiences others as overwhelming him/her with their demands. 'You're forcing me to take medication.' 'I'm being kept here against my will.' 'You're always telling me what I can and can't do.' 'My mother used to lock me in my room.'

Item 10.   The patient repeatedly perceives others as *manipulating and exploiting.* 'You're making me ill so that you can feel better.' 'You're always making me do things.' 'You want me to have a breakdown so you can be right about keeping me here.' 'Nurses are always getting us to cook and clean for them.'

Perspective B

Item 9.   The patient repeatedly perceives himself as *domineering and imposing on others.* 'I can throw my weight around.' 'I can make people listen to me.' 'It's my way or the highway.' 'I'm the boss of this ward; all the patients listen to me.'

Item 10.   The patient repeatedly perceives him/herself as *manipulating and exploiting.* 'I know how to get my needs met.' 'I can force them to do what I want.' 'I'll get my solicitor onto you.' 'I know what my human rights are.' 'I had to constantly tell my mother what to do because she was hopeless.'

Perspective C

Item 9.   Others repeatedly experience the patient as *domineering and imposing.* Others experience the patient as making intrusive take-over attempts so that it is hard to be separate. Others feel intruded into to the extent that the patient evokes feelings or states of mind that really belong to him/her. 'I feel completely taken over.' 'When I go home it's hard to switch off.' 'I'm not allowed to get on with other things.'

Item 10.   Others repeatedly experience the patient as *manipulating and exploiting.* 'He's always targeting vulnerable patients.' 'She watches to see when we're off guard.' 'I know he's going to get others to kick off.'

Perspective D

Item 9.   Others repeatedly experience themselves as *domineering and imposing.* 'I always end up feeling like I'm ordering him around.' 'He makes me want to get heavy with him.'

Item 10.   Others repeatedly experience themselves as *manipulating and exploiting.* 'If you don't do this we're going to re-think your leave.' 'Clean your room ... or else ... '

Blaming

This cluster concerns blame, where the underlying motive is disclaiming and disowning any responsibility and guilt. Unbearable feelings of shame over limitations, inadequacies or failures are consequently perceived as caused by others. This sense of blame

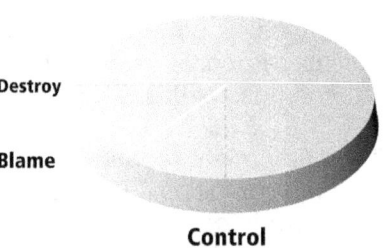

may become so distorted that even harmless statements become accusations. At its more attacking extreme (item 12), one feels humiliated, undermined and criticised by others.

Perspective A

Item 11.   The patient repeatedly perceives others so that they are *accusing* him/her. 'They blamed me for everything that went wrong at home.' 'Everything's always my fault on this ward.' 'My mother blamed me for my father's death.' 'The nurses always want to search me.' 'They're accusing me of being mentally ill.'

Item 12.   The patient repeatedly perceives others so that they are *putting down and humiliating* him/her. 'My family won't visit me – they say hospital is for losers.' 'You're always telling me I can't look after myself.' 'My teachers told me I'd never come to anything.' 'My dad always beat me when he pissed the bed.'

Perspective B

Item 11.   The patient repeatedly experiences him/herself *accusing* others. 'It's all their fault. If they hadn't have pushed me I wouldn't be in this mess.' 'I'll always blame them for stitching me up.' 'I accuse them but they ignore me.'

Item 12.   The patient repeatedly experiences him/herself as *putting down and humiliating others*. 'They're all a load of losers in this place. I wouldn't have anything to do with them on the outs.' 'I need to keep my doctor in his place.' 'I needed to teach my girlfriend a lesson.' 'I needed to bring them down a peg or two.'

Perspective C

Item 11.   Others repeatedly perceive the patient as *accusing* others 'He's always telling us it's all our fault.' 'He blames the nurses for winding him up so that he lost his leave.' 'He tells us we're deliberately trying to make him ill.'

Item 12.   Others repeatedly perceive the patient as *putting down and humiliating them*. 'He's always telling us we haven't got a clue - everything we offer is crap.' 'She's always telling me my group is rubbish in front of the nurses.' 'He's offensive and racially abusive to staff on the ward.'

Perspective D

Item 11.   Others repeatedly experience themselves as if they are *accusing the patient*. 'There's nothing mentally wrong with her – it's all behaviour.' 'When he was a boy he drove his mother to drink.' 'He's always been difficult, he only ever thought of himself.'

Item 12.   Others repeatedly experience themselves as *putting down and humiliating him/her*. 'He's just a nasty piece of work. He's wasting all of our time.' 'Sometimes I feel like I just want to put him in his place.'

Destroying

This cluster concerns active physical attacks or being the victim of such as well as more subtle attacks on care, denigrating the role and capacity of professionals. The patient may spurn offers of help and care, experiencing treatment as an attack, e.g. medication as poison, which leads to counter attack by the patient to eliminate the threat.

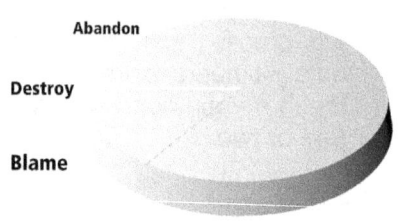

Perspective A.

Item 13.   The patient repeatedly perceives others so that they are *intimidating and attacking him/her.* 'You're coming into my room and raping me every night.' 'You're killing me with this medication.' 'You're trying to make me ill so you can keep me here forever.' 'The nurses are deliberately sabotaging me.'

Item 14.   The patient repeatedly perceives others so that they are *rejecting and excluding him/her.* 'You just want to get rid of me.' 'My mother never liked me she was always a selfish cow.' 'My brothers always ganged up on me and bullied me.' 'I was always the black sheep of the family.'

Perspective B

Item 13.   The patient repeatedly experiences him/herself as *intimidating and attacking.* 'If they come after me again, I'll kill them, I don't care.' 'I meant to kill her, I'm only sorry she survived.' 'You can't be weak in these places – you have to fight to show them know who's on top.' 'Even when my girlfriend was begging me to stop I wanted to smash her face.'

Item 14.   The patient repeatedly experiences him/herself as *rejecting and excluding*. 'I don't care; they don't mean nothing to me.' 'You're dead to me.' 'If somebody crosses me they don't get another chance.'

Perspective C

Item 13.   Others regularly perceive the patient as *intimidating and attacking them*. 'He's terrifying when he gets worked up.' 'He's assaulted lots of staff.' 'I'm always watching my back on my shift.' 'He frightens me.' 'I'm worried he's going to hit me.'

Item 14.   Others regularly perceive the patient as *rejecting and excluding them*. 'She never wants to see me when I come to visit.' 'He's refusing to come to his sessions.' 'He's turning other patients against me.'

Perspective D

Item 13.   Others repeatedly experience themselves as if they are *intimidating and attacking the patient*. 'I'll show him who's boss.' 'He needs to be kept in his place – I'll show him I'm not afraid of him.' 'As a child what he needed was a good hiding.'

Item 14.   Others repeatedly experience themselves as if they are *rejecting and excluding the patient*. 'I want to push him away.' 'He doesn't belong here – we should send him back to prison.' 'He's not my son; I don't want anything to do with him.'

Abandoning

This cluster is relevant for patients with early experiences of loss who can feel easily ignored when staff changes occur or when there are staff holidays leading to the absence of important figures. These patients may also feel loss when they are being discharged from hospital,

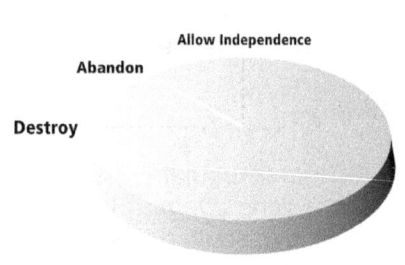

experiencing this as abandonment which can result in attempts to be re-admitted. This can create a counter-response in staff where one realises it has been difficult to keep the patient in mind, or that one has been ignoring their real needs. They may also feel overlooked when they perceive responses as un-empathic or lacking in understanding.

Perspective A

Item 15. The patient repeatedly perceives others so that they are *abandoning him/her.* 'You lot all want to get rid of me.' 'Nobody cares about you in this place.' 'My mother put me in care when I was 12.' 'My dad left after I was born. I've never met him.' 'My doctors never support me in my tribunal.'

Item 16. The patient repeatedly perceives others so that they are *ignoring him/her.* 'Nobody wants to listen.' 'You never pay attention to me.' 'The nurses just ignore you when you bang on the door.' 'I keep telling you the medication is making me feel ill but you don't do anything about it.' 'My teacher didn't even know my name.'

Perspective B

Item 15. The patient repeatedly experiences him/herself as *abandoning others.* 'They can rot in hell for all I care.' 'I couldn't look after my kids – the drugs were all I cared about.' 'Even if my girlfriend was begging me for help I'd walk away.'

Item 16.   The patient repeatedly experiences him/herself as *ignoring others*. 'I keep myself to myself.' 'I don't want to get involved in anything on the ward.' 'You're not worth listening to.' 'When the nurses tell me to clean up my room, I just tune them out.'

Perspective C

Item 15.   Others repeatedly perceive the patient as *abandoning them*. 'He's never been a father to his children.' 'She left all her kids in care.' 'He engaged really well in therapy and then suddenly broke off all contact. Now he blanks me on the grounds.'

Item 16.   Others repeatedly perceive the patient as *ignoring them*. 'He doesn't take any notice of what we say.' 'He never wants to see me when I come to visit.' 'She acts like I don't exist.' 'We keep telling him to take his medication but he won't listen.'

Perspective D

Item 15.   Others repeatedly experience themselves as if they are *abandoning him/her*. 'Sometimes I wish we could just get rid of him.' 'Maybe she'd be better off somewhere else. We haven't got the resources to look after her.' 'I gave him plenty of chances – he's on his own now.'

Item 16.   Others repeatedly experience themselves as if they are *ignoring him/her*. 'You just have to ignore her when she keeps coming to the nursing station.' 'I have to tell him to go away.' 'He's so quiet; I forget he's a patient on the ward.' 'She often slips my mind.'

Asserting

This cluster concerns assertion, where the rigidity of position is such that the individual cannot compromise, adjust or consider alternatives to his/ her own view. The underlying premise is that one's feelings or convictions are obviously the only true

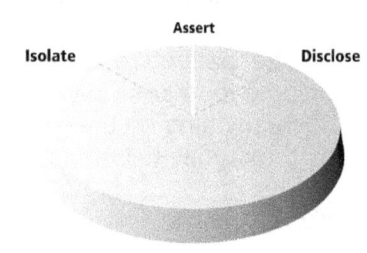

ones. The patient's claims may also take on an antagonistic flavour when the defiance becomes overt, leading to entrenched interpersonal conflicts.

Perspective A

Item 17.   The patient repeatedly perceives others as *defying and opposing him/her.* 'They never let me do what I want, they're always trying to stop me.' 'My mother blocked me at every turn.' 'The doctors like saying no.' 'The nurses just want to stand in your way.'

Item 18.   The patient repeatedly perceives others as *insisting on their position.* 'They never listen to me.' 'You always want your own way.' 'My mother doesn't see my point of view.' 'You keep telling me I've got a mental illness but there's nothing wrong with me.'

Perspective B

Item 17.   The patient repeatedly experiences him/herself as *defying and opposing.* 'I'll never let them win.' 'I'll fight them to the end.' 'I have to make sure that she never gets her way.'

Item 18.   The patient repeatedly experiences him/herself as *insisting on his/her own position.* 'I'm going to stick to my guns.' 'I'm not going to back down over this.' 'I'm right and I know they're wrong.'

Perspective C

Item 17. Others repeatedly perceive the patient as *defying and opposing*. 'He constantly argues with us.' 'It's a battle of wills.' 'She says one thing in the ward round and then does something else.' 'When he was growing up he never let me be a proper mother to him.' 'He started to rebel and then just went off the rails.'

Item 18. Others repeatedly perceive the patient *as insisting on his/her own position*. 'She can't see things from our point of view.' 'He thinks he's right and won't listen to anyone else.' 'He could never see what I was trying to do for him.'

Perspective D

Item 17. Others repeatedly experience themselves as *defying and opposing*. 'It becomes a battle of wills with him.' 'I find I get drawn into endless arguments.' 'We always seem to get into a deadlock.' 'When he was a boy I had to stand my ground otherwise he'd walk all over me.' 'I need to make her see that she's wrong.'

Item 18. Others repeatedly experience themselves as *insisting on their own position*. 'I'm not going to budge.' 'She needs to know we mean business.' 'If I give him an inch he'll take a mile.'

Disclosing

This cluster concerns premature and inappropriate revealing of information about self and/ or others. For example in a first group meeting, the patient gives a long and detailed account in a graphic manner of intimate, sexual or violent relationships including specific aspects of

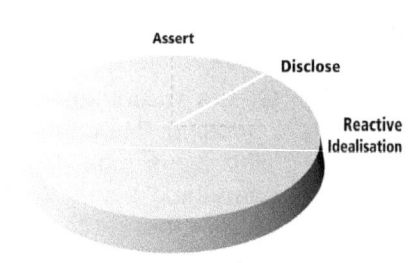

the index offence. The patient may communicate repeatedly and under an apparently irresistible pressure which may have the function of stirring up an emotional response in the listener. For example a professional was regularly approached before the end of the shift by a patient who had killed her children; she said she needed to talk and proceeded to unload gruesome details of what she had done including the exact responses of each child, leaving the staff member over-whelmed.

Perspective A

Item 19.   The patient regularly perceives others as *over revealing and intrusive.* 'My mother always told me things I didn't want to hear.' 'You nurses make it obvious when you've got problems at home.' 'Don't take your bad mood out on me.' 'My psychologist wears tops that show off her breasts.'

Item 20.   The patient regularly perceives others as *pouring out concerns and anxieties.* 'The other patients turn to me for help. I'm a shoulder to cry on.' 'Sometimes the nurses confide in me.' 'My therapist told me he was having problems with his kids too.' 'People like to turn to me with their problems.'

Perspective B

Item 19   The patient regularly experiences him/herself as *over revealing and intrusive.* 'Sometimes I end up saying too much and affecting people.' 'You don't want to hear about my darkest times.' 'Every time I confided in my mother I wound her up.'

Item 20.   The patient regularly experiences him/herself as *pouring out concerns and anxieties.* 'Once I start I can't stop.' 'I end up spilling my guts to anyone who'll listen.' 'The nurses see you at your worst.'

Perspective C

Item 19.   Others regularly perceive the patient as *over revealing and intrusive.* 'He goes into too much detail about his private life.' 'I feel disturbed hearing about his offences.' 'She always ends up telling me things I don't want to hear.'

Item 20.   Others regularly perceive the patient as *pouring out concerns and anxieties.* 'When he talks it's like a flood bursting its banks.' 'Once she starts she doesn't stop.' 'I can never get away once he sees me.' 'When he was a child he worried about everything.'

Perspective D

Item 19.   Others regularly experience themselves as *over-revealing and intrusive.* 'When I'm with him I say things I don't really want to say.' 'She always makes me feel like I've said too much about my personal/ professional life.' 'I end up telling him all the problems at home when I come to visit, even though I know it upsets him.' (Family member.)

Item 20.   Others regularly experience themselves as *pouring out concerns and anxieties.* 'After my primary nurse sessions with her I'm so anxious I end up talking to my husband about it.' 'She makes me feel extremely anxious about what she might do.' 'It's hard to keep it all in.' 'I feel I can tell him anything.'

Reactive Idealising

This cluster indicates a difficulty in recognising negative aspects of the other, often as a response to an idealised view the patient may have of him/herself which can affect particular relationships. Patients regularly single out individual members of staff for the communication of

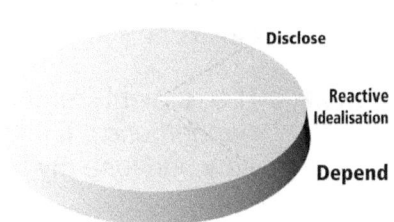

'privileged information'; the recipient is informed that he or she is the only one to whom this information has ever been confided. Reactive idealisation occurs when, in response, the recipient over estimates the patient's point of view including the importance of the staff member. The staff member may, for example, over acknowledge the role of circumstances which brought about the patient behaving in a destructive manner while under emphasising or ignoring completely the patient's own contribution. Alternatively the patient may be seen as highly gifted or superior in some area. In the extreme, this can lead to boundary breaking, for example a staff member taking a patient to his/ her own flat, revealing personal information or having a sexual relationship.

Perspective A

Item 21.   The patient regularly perceives others as *over involved.* 'You understand me completely.' 'I can only talk to S---- as my primary nurse, she's the only one who's behind me all the time.' 'They never let me down.'

Item 22.   The patient regularly perceives others as *over sympathetic.* 'You're the only one who's ever really cared about me.' 'I've never been able to share this with anyone else.' 'I know you'd never judge me.'

Perspective B

Item 21. The patient regularly experiences him/herself as *over involved*. 'I get sucked into other people's problems too easily.'

Item 22. The patient regularly experiences him/herself as *over sympathetic*. 'I always see the other person's point of view and lose my own.'

Perspective C

Item 21. Others regularly perceive the patient as *over involved*.
'He always gets in the thick of things on the ward.' 'He sees someone once and falls in love.' 'He always gets involved in other people's problems.' 'He's a charmer.'

Item 22. Others regularly perceive the patient as *over sympathetic*
'She's always making allowances for others.' 'He can't see how much the other patients are using him.'

Perspective D

Item 21. Others regularly experience themselves as *over involved*
'I end up feeling like I'm the only one who can really help him.' 'I think he just needs love and understanding.' 'I'm the only one he trusts. His secret's safe with me.'

Item 22. Others regularly experience themselves as *over sympathetic*. 'He just got in with the wrong crowd.' 'He didn't do anything wrong.' 'I don't want to get her into trouble.'

Depending

This cluster concerns dependency, expressed either openly or covertly. Most patients in fact have difficulty acknowledging and expressing their feelings of dependency on others, although some are able to communicate their needs openly. Despite the 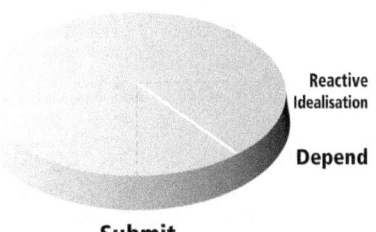 predominant denial, however, dependency is often demonstrated in the patient's behaviour. For example, a regular pattern consists of making good progress over a period of time, leading to optimistic planning for discharge which collapses suddenly when, just as separation from the institution is looming, the patient de-compensates or re-offends, setting the treatment back to square one. The enormous underlying dependency on the structure or physical setting emerges full force. Systematic clinging can also indicate a difficulty in facing dependency, as the recipient needs to be constantly badgered and exhausted: dependency can only be imagined on a person who, in fantasy, has limitless supplies.

Perspective A

Item 23.   The patient regularly perceives others so that they are *over relying on him/her.* 'People are always leaning on me and making me do things I don't want to do.' 'My mum needs me to pay her bills.'

Item 24.   The patient regularly perceives others so that they are *draining him/her.* 'You people are never satisfied. What do you want from me?' 'I can't do anymore.' 'The other patients are always asking me for money or cigarettes.'

Perspective B

Item 23.　The patient regularly experiences him/herself so that he/she is *over relying*. 'People aren't to be trusted, they always let you down.' 'I never learn my lesson, I always get hurt.' 'I don't know what I'd do without my mother.'

Item 24.　The patient regularly experiences him/herself so that he/she is *draining*. The patient may express some paranoid ideation about the malignant effect he fears he may have on others 'I didn't think you'd want to come back and see me anymore.' 'If I tell you what I'm really like inside, you'll not be able to sleep at night.' 'You can't wait to get rid of me'.

Perspective C

Item 23.　Others regularly perceive the patient as *over relying on them*. 'He's always needing us to reassure him.' 'He constantly demands time to sit and talk about his medication/tribunal/side-effects.' 'He always wants me to go out and buy his tobacco and fizzy drinks.'

Item 24.　Others regularly perceive the patient as *draining them*. 'It's impossible to get away.' 'I can't give her any more time and attention.' 'I don't feel I've got anything left to give.' 'I can't do this anymore.'

Perspective D

Item 23.　Others regularly experience themselves as *over relying* on the patient. 'I always look to him to chair the community meeting.' 'He always pays for a taxi when we go on a home visit.' 'He always supports the staff in conflicts with patients.'

Item 24.　Others regularly experience themselves as *draining the patient*. 'I find myself making lots of demands on him.' ''I realise I expect a lot from her.'

Submitting

This cluster concerns submission and compliance, whether used in a placatory way in order to avoid conflict, or used in a more manipulative way in order to play the system or tick the right boxes. When progress and freedom of movement are at the discretion

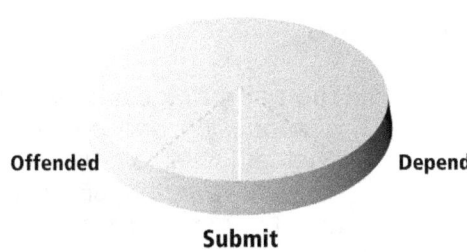

Offended   **Depend**

Submit

of staff, patients often adapt by complying with rules, procedures and treatments in order to advance towards discharge. Staff, faced with the intrinsic threats and anxieties working in secure settings may placate and appease patients to establish a 'peaceful life' together. Reducing doses of medication prematurely; turning a blind eye to minor and not so minor offences; or succumbing to the patient's refusal to engage by giving up – all are common patterns. Staff may also find themselves submitting to patients' constant demands and relentless requests, appeasing the patient when they are afraid of the consequences of refusing demands. At its most extreme both staff and patients may find themselves giving up in despair. This is sometimes evident in resigned and defeated behaviour and the experience that one has simply run out of resources.

Perspective A

Item 23. The patient regularly perceives others so that they are *appeasing and complying with him/her.* 'You doctors haven't got a clue. You're just going along with what I want.' 'My mother only tells me what I want to hear.' 'You're only saying that to get me to go to groups.' 'The patients here only pretend to like you to get something from you.'

Item 24. The patient regularly perceives others so that they are *giving up in despair.* 'You don't care about me. You're leaving me here to rot.' 'My OT doesn't come to see me anymore, she gave up on me.' 'My mother put me in a home when I was 12, she didn't know what else to do with me.'

Perspective B

Item 23.   The patient regularly experiences him/herself so that he/she is *appeasing and complying.* 'I have to do what they say whether I want to or not.' 'I just go along with what they want.' 'To survive in this place, you have to keep your head down and do what you're told.'

Item 24.   The patient regularly experiences him/herself so that he/she is *giving up in despair.* 'I just don't care what happens to me anymore.' 'I don't care if I live or die.'

Perspective C

Item 23.   Others regularly perceive the patient as *appeasing and complying.* 'He only tells us what we want to hear.' 'She goes along with our plans but then doesn't say what she really wants.'

Item 24.   Others regularly perceive the patient as *giving up in despair* 'She's in bed all day, we can't engage her in anything.' 'He's stopped looking after himself.'

Perspective D

Item 23.   Others regularly experience themselves as *appeasing and complying.* 'I feel I have to give in to his demands otherwise he becomes threatening to the staff.' 'Sometimes it's easier to say yes, even though I know I shouldn't.'

Item 24.   Others regularly experience themselves as *giving up in despair.* 'We've reached the end of the road with him. There's nothing else we can do.' 'We never see her in the ward rounds, there's no point.' 'He's happy to sleep all day and we just let him.'

Taking Offence

This cluster concerns taking offence in either a narcissistic or paranoid form. For example there may be over sensitivity to a different viewpoint, experienced as hostile or as offensive to the patient's dignity, which in turn triggers wounded vulnerability. Or, the

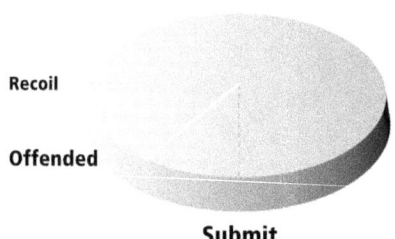

patient evinces sudden feelings of betrayal by the other who has responded in a manner which does not seem to acknowledge an intention or need. Similarly, staff may feel enormously betrayed and cheated when patients do not respond to their care and concern. For example patients who abscond in spite of giving reassurance to staff may leave the latter feeling undermined and attacked. The more paranoid version is when the different view is not so much a slight as a threat, and the response becomes forceful, uncompromising indignation which in the extreme becomes a torrent of self assertion.

Perspective A

Item 23.   The patient regularly perceives others so that they are *indignant and self-justifying.* 'You doctors don't want to admit you're in the wrong – you all cover for each other.' 'He never apologised for hitting me – he's too busy getting the nurses on his side.' 'She could never listen to me – all she's interested in doing is defending her own corner.'

Item 24.   The patient regularly perceives others so that they are *hurt and touchy.* 'I can't say anything to you people - you don't know how to take a joke.' 'I only gave him a little push.' 'My mother always over-reacted.'

Perspective B

Item 23.   The patient regularly experiences him/herself so that he/she is *indignant and self-justifying.* 'I didn't mean to do it but she forced me.' 'He was asking for it.' 'There's nothing wrong with me, I was just acting out of self defence.'

Item 24.   The patient regularly experiences him/herself so that he/she is *hurt and touchy.* 'I can't believe you ignored me. I only wanted you to talk to me for five minutes.' 'It's just a small gift, why won't you take it?'

Perspective C

Item 23.   Others regularly perceive the patient as *indignant and self-justifying.* 'I had to get the knife, you lot weren't listening to me.' 'There's nothing wrong with me – you're the mad ones.' 'He was asking for it.'

Item 24.   Others regularly perceive the patient as *hurt and touchy.* 'You can't say anything right.' 'He chooses to take offence at everything you say.' 'He was always a sensitive boy – I couldn't do a thing right.'

Perspective D

Item 23.   Others regularly experience themselves as *indignant and self-justifying.* 'She makes me feel I have to explain myself.' 'I always feel as if I'm being accused of malpractice.' 'He always makes me feel defensive and caught on the back foot.'

Item 24.   Others regularly experience themselves as *hurt and touchy.* 'I really trusted him that time. I can't believe he let me down.' 'I feel as if he's betrayed my trust.' 'I can't help but take it personally when he insults me.'

Recoiling

This cluster concerns a fight/flight reaction to over-whelming stress or threat with either a schizoid or autistic flavour. Attacking attitudes and behaviour oscilate suddenly with predominant avoidance or fear. Patients may avoid attendance of particular

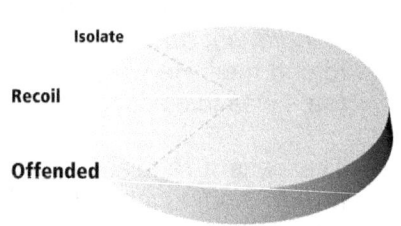

meetings, for example, a ward community meeting or treatment groups. In some cases staff may find themselves evading issues or activities. For example a staff member serving a meal was accused by a patient of attempting to punch him, and the staff member responded by saying he would never serve meals again, even though the accusation was totally untrue. 'Human rights' allegations and other accusations, let alone the potential for physical assaults, can elicit flight in staff. Revulsion over contact with a paedophile or a child killer may also lead to psychological flight, as well as forgetting about the patient's index offence. When patients attribute their own disgusting parts to another, the latter can be at risk of attack: flight quickly alternating with fight.

Perspective A

Item 23. The patient regularly perceives others so that they are *running away.* 'You're always avoiding me.' 'You've never got time to listen to me properly.' 'My teachers always went the other way when they saw me.'

Item 24. The patient regularly perceives others so that they are *showing disgust.* 'What's wrong with your fucking face?' 'She's always turning her nose up at me.' 'My girlfriend thought I was a monster.' 'The nurses hate looking at my scars.'

Perspective B

Item 23.   The patient regularly experiences him/herself so that he/she is *running away.* 'I just decided not to come back from my leave and get some drugs.' 'I just want to get away from this place.' 'I want to get off my face.' 'I can't think about that stuff right now.'

Item 24.   The patient regularly experiences him/herself so that he/she is *showing disgust.* 'I can't stand you useless staff - you're rubbish.' 'This place is full of losers and wasters.' 'I'm better than these animals.'

Perspective C

Item 23.   Others regularly perceive the patient as *running away.* Patient frequently absconds or walks out of meetings/ sessions. 'He always leaves when things get tough.' 'My son always ran out on me when his father came home drunk.' 'At the first sign of trouble he was off.' 'He always takes drugs when things get difficult.'

Item 24.   Others regularly perceive the patient as *showing disgust.* Patient is abusive and offensive towards staff/other patients. 'Get your dirty hands off me.' 'This place is full of animals.'

Perspective D

Item 23.   Others regularly experience themselves as *running away.* 'I just want to stay in the nursing station.' 'I always want to avoid her if I can.' 'I make sure we don't get into anything too difficult in our therapy sessions.' 'I don't want to have to deal with him at the moment.'

Item 24.   Others regularly experience themselves as *showing disgust.*

'His smell often makes it hard to be in the same room as him.' 'I don't want to have to deal with her cutting.' 'He makes me want to throw up.'

Isolating

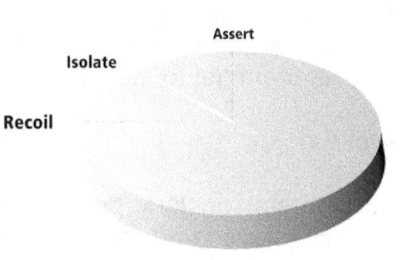

The patient who finds refuge in his or her own room or the staff member who walls his or herself off in the nursing office are examples of how any contact can come to be perceived as dangerous. Or, the patient may interact but gives a constant impression of having an impermeable shield in place behind which he/she vets every communication from the other and responds with carefully packaged, laconic or hostile ('mind your own business') attitudes. Staff as well can adopt formal 'professional' roles which allow little emotional contact or interaction.

Perspective A

Item 23.　The patient regularly perceives others so that they are *cutting off contact*. 'They don't want to have anything to do with me.' 'My mother just goes on holiday without telling me.' 'My psychologist suddenly stopped seeing me.'

Item 24.　The patient regularly perceives others so that they are *keeping up a barrier*. 'I never feel as if you're listening to me.' 'You've made up your mind, you're not interested in what I've got to say.' 'You never tell me anything about yourself.'

Perspective B

Item 23.　The patient regularly experiences him/herself so that he/she is *cutting off contact*. 'I don't want anything to do with them.' 'I don't want to see you anymore.' 'I'm not allowing my family to visit me.'

Item 24.   The patient regularly experiences him/herself so that he/ she is *keeping up a barrier.* 'I just let them get on with it.' 'I don't let them get to me.' 'I'm not telling them anything.'

Perspective C

Item 23.   Others regularly perceive the patient as *cutting off contact.* 'He stays in his room all day.' 'She refuses to talk to us.' 'She walks right past me when I try to say hello.'

Item 24.   Others regularly perceive the patient as *keeping up a barrier.* 'He won't tell us what he's feeling.' 'He just goes along with what we want but there's no real contact.' 'It's hard to know what he's really thinking.' 'We don't really know her, even though we've been treating her for years.'

Perspective D

Item 23.   Others regularly experience themselves as *cutting off contact.* 'I can't stand to be in the same room as her.' 'I can't work with him anymore.' 'I went on leave and realised I'd forgot to tell him.'

Item 24.   Others regularly experience themselves as *keeping up a barrier.* 'I don't like to interact with him.' 'I don't like to get too close to her.' 'I never like him to know what I'm thinking.'

Bibliography

AMERICAN PSYCHIATRIC ASSOCIATION (2000) *Diagnostic and Statistical Manual of Mental Disorders-IV-TR.* APA Press.

ALLPORT, G.W. (1937) *Personality: A Psychological Interpretation.* Henry Holt.

BATEMAN, A.W. and FONAGY P. (2004) *Psychotherapy for Borderline Personality Disorder: Mentalization-Based Treatment:* Oxford University Press.

BENJAMIN, L.S. (1996) *Interpersonal Diagnosis and Treatment of Personality Disorders*: Guilford Press.

BIRTCHNELL, J. (1993) *How Humans Relate: A New Interpersonal Theory.* Praeger.

BLACKBURN, R. (1998) Criminality and the interpersonal circle in mentally disordered patients. *Criminal Justice and Behaviour* 25: 155 – 76.

BOWLBY, J. (1969) *Attachment and Loss. Vol 1: Attachment.* Hogarth Press/Institute of Psychoanalysis.

CASEMENT, P. (1985) *On Learning from the Patient.* Routledge, London.

DAHLBENDER, R.W., RUDOLF, G. & OPD Task Force (2006). Psychic structure and mental functioning: Current research on the reliable measurement and clinical validity of Operationalized Psychodynamic Diagnostics (OPD) System. In PDM Task Force (Ed.), *Psychodynamic Diagnostic Manual,* 615-662. Silver Spring: Alliance of Psychoanalytic Organizations.

FREUD, S. (1950) *'Why War?' Collected papers, Vol 5* (trans J Strachey). Hogarth Press.

GREENBURG, J.R. and MITCHELL, S.A. (1983) *Object Relations in Psychoanalytic Theory.* Harvard University Press.

GROSS, S., STASCH, M., SCHMAL, H., HILLENBBRAND E. and CIERPKA, M. (2007): Changes in the mental representations of relational behaviour in depressive patients. *Psychotherapy Research* 17(5): 522-534.

GUTTMAN, L.C. (1996) Order Analysis of Correlation Matrices. In R.B. CATTELL (Ed.), *Handbook of Multivariate Experimental Psychology*, 439-58. Rand McNally.

HOFFMAN, I.Z. and GILL, M.M. (1988) A scheme for coding the patient's experience of the relationship with the therapist (PERT): some applications, extensions and comparisons. In H. DAHI , H. KACHELE and H. THOMÄ (Eds.), *Psychoanalytic Process Research Strategies*, 67 – 98. Springer-Verlag.

HOROWITZ, M. (1991) *Personal Schemas and Maladaptive Interpersonal Behaviour.* University of Chicago Press.

KIRTCHUK, G., REISS, D. and GORDON, J. (2008) Interpersonal dynamics in the everyday practice of a forensic unit. In J. GORDON and G. KIRTCHUK (Eds.) *Psychic Assaults and Frightened Clinicians: Countertransference in Forensic Settings,* 97-112. Karnac.

LEARY, T. (1957) *Interpersonal Diagnosis of Personality: A functional Theory and Methodology for Personality Evaluation.* Ronald Press.

LUBORSKY, L. and CRITS-CHRISTOPH, P. (1990) *Understanding Transference.* Basic Books.

MURRAY, H.A. (1938) *Explorations in Personality.* Oxford University Press.

OPD Task Force (2001) *Operationalized Psychodynamic Diagnostics: Foundations and Manual.* Hogrefe & Huber.

OPD Task Force (2008) *Operationalized Psychodynamic Diagnosis OPD-2: Manual of Diagnosis and Treatment Planning.* Hogrefe & Huber.

RACKER, H. (1968) *Transference and Countertransference.* International University Press.

SCHAEFER, E.S. (1965) Configuration analysis of children's reports of parent behaviour. *Journal of Consulting Psychology* 29: 552 – 7.

SCHORE, A. (2001) Effects of a secure attachment relationship on right brain development, affect regulation and infant mental health. *Infant Mental Health Journal* 22(1-2): 7-66.

STASCH, M. (2004) Interpersonal tuning in inpatient psychotherapy. A clinical approach based on the Operationalized Psychodynamic Diagnostics (OPD). 34th annual meeting of the Society for Psychotherapy Research. Weimar.

STASCH, M., CIERPKA, M., HILLENBRAND, E. and SCHMAL, H. (2002) Assessing re-enactment in inpatient psychodynamic therapy. *Psychotherapy Research* 12 (3): 355-368.

STRUPP, H. and BINDER J. (1984) *Psychotherapy in a New Key. A Guide to Time Limited Dynamic Psychotherapy.* Basic Books.

SULLIVAN, H.S. (1953) *The Interpersonal Theory of Psychiatry.* Norton.

VON DER TANN, M., CIERPKA, M., GRANDE, T. and STASCH, M. (2007) The Operationalised Psychodynamic Diagnostics System: Clinical Relevance, Reliability and Validity. *Psychopathology* 40:209-220.

WEISS, J. and SAMPSON H. (1986) *The Psychoanalytic Process: Theory, Clinical Observation and Empirical Research.* Guilford Press.

WORLD HEALTH ORGANIZATION (WHO) (1994) *International Statistical Classification of Diseases and Related Health Problems - 10*. WHO Press.

Appendix

Interpersonal Dynamics Worksheet

Identifying Information

Patient

Name:

ID:

Date of Birth:

Ethnicity:

Date of Admission:

Referral

Date of Referral:

Referral Source:

Difficulties that the team would like to address in this consultation:
(e.g. is the treatment 'stuck' in any way, does the team disagree about the patient, difficulty engaging the patient, does the patient sabotage any progress?)

Form adapted from HCR-20 – Assessing Risk for Violence, Version 2. By C. Webster, K. Douglas, D. Eaves and S. Hart. Mental Health Law and Policy Institute, British Columbia, 1997.

Further Information

Details

Time:
Place:

Source of Information Reviewed

☐ Patient Interview

☐ Patient Medical Records

☐ Staff Consultation: Team discussion

☐ Collateral Interviews

☐ Other

Summary of Relationship History

Significant Others in Early Childhood :

Adolescence:

Adulthood:

Index Offence:

What was the relationship of the victim to the patient: (e.g. was victim known to the patient?)

Previous offending history. Response generated by health and legal systems (e.g. sectioned/bailed/no consequence)

Current relationships: a) with staff

                                       b) other patients

                                       c) significant others

PERSPECTIVE	EVIDENCE
A) Time and again the Patient experiences staff and others as ...	
1.	
2	
3	
4	
B) Time and again the Patient experiences him or herself as ...	
1.	
2.	
3.	
4.	
C) Staff and others time and again experience the patient as ...	
1.	
2.	
3.	
4.	
D) Staff and others time and again experience themselves	
1.	
2.	
3.	
4.	

Brief Formulation (according to four perspective menu):

_____ ⇒ _____ b)
_____   _____
_____   _____
⇑ _____   _____ ⇓
_____   _____
_____   _____
d) _____ ⇐ _____ c)

Expanded Formulation (including a longer narrative and hypothesis):

Conclusions and recommendations with regard to risk:

Conclusions and recommendations for CPA care plans

INTERPERSONAL DYNAMICS

A. The patient repeatedly perceives others so that they are ...
    D. Others, the staff included, regularly experience themselves as

B. The patient regularly experiences himself/herself as
    C. Others, the staff included, regularly perceive the patient as

A (D)	No.	Description (D)	CLUSTERS	No.	Description (C)	B
☐ ☐	1	Treating him/her as self-sufficiently independent	Allowing Independence	1	Self-sufficiently independent	☐ ☐
☐ ☐	2	Behaving as though he/she knows best		2	Behaving as though he/she knows best	
☐ ☐	3	Recognising and accepting	Affirming	3	Recognising and accepting	☐ ☐
☐ ☐	4	Approving and admiring		4	Approving and admiring	
☐ ☐	5	Treating him/her as special	Idealising	5	Treating him/herself as special	☐ ☐
☐ ☐	6	Idealising him/her		6	Idealising	
☐ ☐	7	Attending to and caring for him/her in every way	Protecting	7	Attending to and caring	☐ ☐
☐ ☐	8	Instructing and patronising him/her		8	Instructing and patronising	
☐ ☐	9	Domineering and imposing on him/ her	Controlling	9	Domineering and imposing	☐ ☐
☐ ☐	10	Manipulating and exploiting		10	Manipulating and exploiting	
☐ ☐	11	Accusing him/her	Blaming	11	Accusing	☐ ☐
☐ ☐	12	Putting down and humiliating him/her		12	Putting down and humiliating	
☐ ☐	13	Intimidating and attacking him/her	Destroying	13	Intimidating and attacking	☐ ☐
☐ ☐	14	Rejecting and excluding him/her		14	Rejecting and excluding	
☐ ☐	15	Deserting him/her	Abandoning	15	Deserting	☐ ☐
☐ ☐	16	Ignoring him/her		16	Ignoring	
☐ ☐	17	Defying and opposing him/her	Asserting	17	Defying and opposing	☐ ☐
☐ ☐	18	Insisting on their position		18	Insisting on his/her position	
☐ ☐	19	Over revealing and intruding	Disclosing	19	Over revealing and intruding	☐ ☐
☐ ☐	20	Pouring out concerns and anxieties		20	Pouring out concerns and anxieties	
☐ ☐	21	Over involved	Reactive Idealising	21	Over involved	☐ ☐
☐ ☐	22	Over sympathetic		22	Over sympathetic	
☐ ☐	23	Over relying on him/her	Depending	23	Over relying on	☐ ☐
☐ ☐	24	Draining him/her		24	Draining	
☐ ☐	25	Appeasing and complying with him/her	Submitting	25	Appeasing and complying with	☐ ☐
☐ ☐	26	Giving up in despair		26	Giving up in despair	
☐ ☐	27	Hurt and touchy	Taking Offence	27	Hurt and touchy	☐ ☐
☐ ☐	28	Indignant and self-justifying		28	Indignant and self-justifying	
☐ ☐	29	Showing Disgust	Recoiling	29	Showing Disgust	☐ ☐
☐ ☐	30	Running away		30	Running away	
☐ ☐	31	Cutting off contact	Isolating	31	Cutting off contact	☐ ☐
☐ ☐	32	Keeping up a barrier		32	Keeping up a barrier	

Form adapted from Operationalized Psychodynamic Diagnostics: Foundations and Manual by OPD Task Force. Hogrefe & Huber, 2001.